FLORA OF TROPICAL EAST AFRICA

PITTOSPORACEAE

G. CUFODONTIS
(University of Vienna)

Evergreen trees, shrubs or climbers with resin-ducts in the bark, rarely spiny. Leaves alternate, often crowded at the ends of the branches, simple, entire, dentate or lobed, ± leathery. Stipules absent. Flowers few to many in terminal and/or axillary panicles or cymes, sometimes in clusters on the old wood or solitary in the axils, regular or rarely slightly irregular, hypogynous, bisexual or functionally (rarely morphologically) unisexual. Sepals 5, free and imbricate at least in bud, or somewhat connate. Petals 5, imbricate in bud, generally free, rarely with slightly connivent claws. Stamens 5, alternate with petals, free or with somewhat connivent filaments; anthers dithecous, introrse, opening by slits or pores. Ovary superior, sessile or shortly stipitate, paracarpous, with 2–5 carpels and parietal placentas, usually unilocular but sometimes 2–5-locular by central contact of placentas; style simple with capitate or somewhat lobed stigma. Fruit a berry or a capsule with generally entire valves. Seeds without an aril but often covered by a viscid resin, rarely dry and winged; testa thin and smooth; endosperm copious and horny; embryo minute.

In addition to the only African genus *Pittosporum*, there are 8 genera, mainly from Australasia and Malesia.

PITTOSPORUM

Soland. in Gaertn., Fruct. 1: 286, t. 59/7 (1788); Pritzel in E. & P. Pf., ed. 2, 18a: 273 (1930); J. Léon. in F.C.B. 2: 574 (1951); Cuf. in F.R. 55: 27 (1952) & in F.W.T.A., ed. 2, 1: 182 (1954) & in Fl. Madag. (1955) & in F.Z. 1: 298 (1960) & in Bol. Soc. Brot., sér. 2, 34: 159 (1960); K.T.S.: 380 (1961); F.F.N.R.: 66 (1962)

Trees or shrubs, never climbing or spiny. Leaves entire or (outside Africa) rarely undulate or sinuate, glabrous or hairy. Inflorescences usually many-flowered subracemose or subumbellate panicles, terminal and/or axillary from the uppermost leaves, but (outside Africa) flowers sometimes solitary or clustered on the old wood. Flowers regular, up to 15 mm. long, of various colours (but only white, greenish or yellow in Africa), sweet-scented, very often, perhaps always, functionally unisexual; ♂ with long filaments, fertile anthers and slender sterile ovary; ♀ with short filaments, reduced sterile anthers and stout fertile ovary. Ovary unilocular, 2(–5)-merous; style ± as long as ovary, with truncate or sublobed stigma, splitting in fruit according to number of carpels; ovules ovoid. Fruit a capsule; valves leathery or woody, yellow or brown, finally suberect, spreading or reflexed. Seeds variously deformed by mutual pressure, 2–4(–many outside Africa) in two rows on each placenta, ripening orange or red, covered with a sticky slow-drying resin.

Up to 200 species extending from Africa and Madagascar to China and Japan, also to Australia, New Zealand and Pacific Is.

The distribution of sexes is difficult to establish, but each inflorescence examined, however, was found to have flowers only of the same sex. The gynoecium of the African species is almost always dimerous and in fact only two cases of trivalvate capsules (together with other bivalvate ones) have been seen among all the African material examined, namely on *Fries* 928, the type of *P. viridiflorum* Sims var. *afrorientale* (Cuf.) Cuf. from Kenya, and on one specimen of *P. viridiflorum* var. *viridiflorum* from South Africa. The number of seeds in capsules of African species seems never to exceed 8, i.e. two pairs on each valve.

P. undulatum Vent. (fig. 1/9), a native of SE. Australia, has been cultivated in Tanganyika at Amani (*Grote* in *Herb. Amani* 6560 & *Greenway* 1617 & 2824). It is also recorded as an escape from cultivation in Kenya around Nairobi, e.g. Kabete, *Mettam* 137. It is easily recognized by its lanceolate glabrous leaves with wavy margins, yellowish flowers up to 12 mm. long and calyx 5–7 mm. long, connate and split on one side, with arcuate subulate pubescent lobes.

NOTE. *P. spathulifolium* Engl. and *P. jaegeri* Engl. are both based on specimens of *Turraea mombassana* C. DC. (*Meliaceae*) from Tanganyika.

Leaves quite sessile or petiole not longer than 2–3 mm.; inflorescences subracemose, few-flowered; flowers up to 14 mm. long; pedicels up to twice as long as flowers . . . . 6. *P. goetzei*

Leaves with petioles always longer and up to ± 20 mm. long; inflorescences racemose or subumbellate panicles, usually many-flowered and if rather few-flowered never subracemose; flowers rarely attaining 10 mm., usually shorter; pedicels often shorter than flowers, never twice as long:

Adult leaves covered beneath by a dense persistent fulvous or ochraceous tomentum; sepals free, ovate, obtuse, imbricate; ripe capsules generally more than 4-seeded, with valves thick and woody at base, thinner upwards, dorsally convex, suberect-divergent; ovary tomentose . 5. *P. lanatum*

Adult leaves glabrous or thinly pubescent and often glabrescent beneath; sepals either connate or if free not imbricate:

Leaves always quite glabrous even when young; vein-reticulation dense, regular, persistently reddish-brown beneath; young twigs, branches of inflorescence and pedicels hispidulous with shining coppery hairs; all parts of flower quite glabrous; sepals rounded, slightly connivent at base; ripe capsule up to 8-seeded, with flat horizontally spreading valves . . . . . . 4. *P. lynesii*

Leaves either ± hairy, at least when young, or if glabrous without reddish-brown reticulation beneath; branches of inflorescence glabrous or hairy but not with shining coppery hairs:

Flowering branches with leaves not evidently crowded at the ends; pubescence of leaves and inflorescence ± spreading, persistent even on midrib and petiole; inflorescences generally slender, conoid; calyx high-connate with short rounded pubescent lobes, at first enveloping the buds and in anthesis split down one side and spathe-like; petals mostly revolute from the middle: capsules more than 4-seeded . 3. *P. spathicalyx*

Flowering branches with leaves mostly crowded at the ends; pubescence of leaves almost always evanescent or absent; calyx of free sepals or if connate not with such short rounded pubescent lobes:

Leaves usually ± acuminate, quite glabrous with age, rarely puberulous at base when young, generally much paler beneath than above; midrib impressed above and sometimes almost hidden in a groove; lateral nerves, at least with age, somewhat prominent beneath and ± conspicuously interconnected by loops, slightly engraved above; reticulation not uniform, rather lax and more so along midrib, ± coloured beneath; branches of inflorescence and pedicels always pubescent; sepals free; ripe capsules never more than 4-seeded, with the valves rather thin, dorsally concave and finally reflexed . . . . 2. *P. mannii* subsp. *ripicola*

Leaves often obovate-spathulate, obtuse or rounded, often glaucescent glabrous or pubescent (at least when young) on midrib and petiole; midrib flat or nearly so above; lateral nerves thin and flat beneath, merging into the reticulation, often protruding above with age; reticulation dense, uniform, with its colour soon disappearing beneath; branches of inflorescence and pedicels glabrous or pubescent; sepals glabrous or pubescent, free or ± high-connate but if so glabrous like the whole plant; ripe capsule 4–8-seeded, with valves not exactly as above . . . . 1. *P. viridiflorum*

1. **P. viridiflorum** *Sims* in Curtis, Bot. Mag. 41, t. 1684 (1814); Hiern, Cat. Afr. Pl. Welw. 1: 41 (1896); Dur. & Schinz, Consp. Fl. Afr. 1 (2): 228 (1898); Exell in J.B. 64, Suppl.: 21 (1926); C.F.A. 1: 87 (1937) & Cuf. in C.F.A. 1: 362 (1951) & in F.R. 55: 41 (1952) & in Fl. Madag., Pittosporacées: 18, fig. 3/1–12 (1955) & in F.Z. 1: 299, t. 54A (1960) & in Bol. Soc. Brot., sér. 2, 34: 164 (1960). Type: illustration in Curtis, Bot. Mag. 41, t. 1684 (1814) of a plant from the Cape of Good Hope cultivated in London

Shrub or tree up to 20 m. high; bole up to 50 cm. in diameter; bark of bole and older twigs pale to dark grey, greyish-brown to brown or rarely blackish, smooth, occasionally peeling off like paper. Leaves mostly crowded at ends of branches, with blade spathulate, obovate or broadly oblanceolate, together with petiole up to 14·5 cm. long, 4·5 cm. broad (average ± 9·5 × 3 cm.), rounded, rarely subtruncate, mostly ± acuminate with blunt tip, narrowed into the up to 20 mm. long petiole, often glaucous-green, paler beneath, quite glabrous or somewhat pubescent on midrib and petiole when young, later ± glabrescent; midrib flat or slightly impressed above, prominent beneath; lateral nerves up to 8 on either side, fine but often somewhat prominent above, merging into the dense uniform reticulation, which is persistently visible and areolate or tessellate above, dark coloured beneath, but usually soon turning pale. Inflorescences terminal short racemose or subumbellate panicles, up to 6 cm. long and broad; branches glabrous to shortly hairy; bracts subulate, up to 3 mm. long, puberulous, caducous. Flowers with strong sweet scent like lemons or *Jasminum*, variously white, cream, greenish, yellowish or rarely golden green, on 3–6–10* mm. long glabrous, puberulous or shortly hairy pedicels. Calyx 1–3(–6) mm. long, glabrous or somewhat puberulous; sepals either free and not imbricate, bluntly ovate-lanceolate, ± 1 mm. broad, or connate into a 1–4·5 mm. long tube (often split down one side) and with irregular lobes. Petals (4·5–)5·5–7–9 mm. long and in upper half 1·5–2–2·5 mm. broad, mostly erect with spreading tips. Fertile stamens 4–4·8–6 mm. long with 1·2–1·8–2·2 mm. long anthers; sterile stamens 2·6–5·5 mm. long with anthers not exceeding 1 mm. in length. Gynoecium 4–5 mm. long if fertile, 4·5–6·5 mm. long if sterile, usually glabrous. Capsules 4–8-seeded; valves of ripe capsule up to 10 mm. in diameter, sometimes longer than broad, plano-convex or dorsally gibbous, spreading or bent slightly upwards.

subsp. **viridiflorum**

Leaves less than 4 times as long as broad; inflorescences 8–many-flowered.

NOTE. In addition to subsp. *viridiflorum*, which ranges from tropical and extratropical Africa at least as far as Madagascar and southern India, two little known subspecies are maintained for the time being. Subsp. *arabicum* Cuf., of which the flowers are still unknown, has been collected only twice (*Schweinfurth* 1715, *Deflers* 408) from a restricted area in tropical Arabia, while subsp. *meianthum* Cuf. is known only from the holotype (*Humbert* 11699) from southern Madagascar. Subsp. *viridiflorum* alone occurs in Africa and is exceedingly variable. While the morphological variation is partly related to factors of distribution and habitat so that some subdivision seems desirable, the system adopted here cannot be regarded as entirely satisfactory and even less definitive. The main handicap is an imperfect knowledge of the correlation between characters of the fruits and flowers, which are very rarely collected together so that reliance must be placed mostly on inductive considerations. All varieties are ± connected by intermediates and not rarely the allotment of a specimen to one or another is unfortunately uncertain.

KEY TO VARIETIES OF SUBSP. VIRIDIFLORUM

Branches of inflorescence, pedicels and calyx ± hairy or pubescent; mature capsules generally (perhaps always in East Africa) 4-seeded; valves of ripe capsule stiff but relatively thin, distinctly gibbous-convex dorsally and usually slightly bent upwards:

- Sepals free from each other or nearly so . . . var. **viridiflorum**
- Sepals ± high-connate and calyx sometimes splitting with age . . . . . . var. **malosanum**

Branches of inflorescence, pedicels and calyx quite glabrous or only sparsely adpressed puberulous; mature capsules (in East Africa at least) always more than 4-seeded; valves of completely ripe capsules thick, nearly flattened dorsally and ± horizontally spreading:

- Sepals free from each other or nearly so . . . var. **afrorientale**

*Wherever three measurements are given the middle one indicates the approximate average.

Sepals ± high-connate or connivent and calyx often
splitting with age . . . . . . var. **kruegeri**

var. **viridiflorum**; Cuf. in Bol. Soc. Brot., sér. 2, 34: 164 (1960)

Adult leaves with breadth-length ratio 1:2–1:3·5, rounded or shortly acuminate, usually glabrescent and glaucescent. Pedicels 3–5·2–7 mm. long, like all branches of inflorescence ± pubescent. Sepals free, 1–2·5 mm. long, pubescent. Petals 4·5–9 × 2–2·5 mm. Ripe capsules mostly (probably always in East Africa) 4-seeded; valves thin, stiff, dorsally gibbous-convex, rarely concave, slightly bent upwards or horizontal. Fig. 1/1, p. 6.

TANGANYIKA. Lushoto District: Maramba, 23 July 1917, *Zimmermann* in *Herb. Amani* 6561! & Monga, 24 Nov. 1916, *Zimmermann* in *Herb. Amani* 6562!; Songea District: 32 km. E. of Songea by R. Mkurira, 8 June 1956, *Milne-Redhead & Taylor* 10725!
DISTR. T3,8; Somali Republic (N.), Malawi, Rhodesia, South Africa (Transvaal, Natal, Cape Province) and Angola, also St. Helena (probably introduced), Madagascar and S. India
HAB. Lowland and upland rain-forest and riverine forest; 900–1200 m.

SYN. *P. floribundum* Wight & Arn., Prodr. Fl. Ind. Or. 1: 154 (1834). Type: India, near Nilghiri, *Wight* 976 (K, holo. !, A,NY, iso.)
P. commutatum Putterl., Syn. Pittosp.: 10 (1839). Types: South Africa, Cape Province, Knysna, *Ecklon* 236 & *Drège* 6181 (W, syn. !)
P. abyssinicum Del. var. *angolensis* Oliv., F.T.A. 1: 124 (1868). Type: Angola, Huila, *Welwitsch* 1034 (K, holo. !, BM,P, iso. !)
[*P. mannii* sensu P.O.A. C: 190 (1895), quoad Benguella, *non* Hook. f.]
[*P. abyssinicum* sensu Engl. in Sitz. Ber. Akad. Berlin 10: 408 (1904); Gilliland in Journ. Ecol. 40: 121–123 (1952) et auct. mult. pro parte, *non* Del.]
P. vosseleri Engl. in E.J. 43: 371 (1909) & in V.E. 3 (1): 851 (1915), pro parte. Types: Tanganyika, E. Usambara Mts., Longuza [Lungusa], *Engler* 416 & W. Usambara Mts., Kwai, *Albers* 267 (both B, syn. †)
P. viridiflorum Sims var. *commutatum* (Putterl.) Engl. in V.E. 3(1): 850 (1915)
P. viridiflorum Sims subsp. *viridiflorum*; Cuf. in F.R. 55: 59, fig. 4/a–c (1952)
P. viridiflorum Sims subsp. *angolense* (Oliv.) Cuf. in F.R. 55: 62, fig. 4/i (1952)
P. viridiflorum Sims subsp. *somalense* Cuf. in F.R. 55: 64 (1952), pro parte, excl. fig. 4/k, l, & E.P.A.: 175 (1954). Type: Somali Republic (N.), Tabah Gap gorge to Erigavo, *Glover & Gilliland* 1115 (K, holo. !)

NOTE. The typical variety occurs mainly in the southern part of the species range, especially in the Cape Province of South Africa and in Rhodesia. From Malawi northwards it becomes very rare and north of Tanganyika only two specimens (*Hildebrandt* 1533, *Glover & Gilliland* 1115) from the eastern part of the Somali Republic (N.) are referable to it. In most of its localities var. *malosanum* is also encountered and not infrequently of transitional form. In Angolan and some South African specimens free and ± connate sepals sometimes appear on the same individual, so there is evidently a close relationship between the two varieties.

P. vosseleri Engl. certainly belongs here, although no type material apparently now exists for comparison.

var. **malosanum** (*Baker*) *Cuf.* in Bol. Soc. Brot., sér. 2, 34: 164 (1960). Type: Malawi, near Zomba, Mt. Malosa, *Whyte* 420 (K, holo. !)

Adult leaves with breadth-length ratio 1:2·4–1:3·8, ± acuminate, sometimes puberulous on midrib and petiole, often dull green above. Pedicels 5–6·7–9 mm. long, like all the branches of inflorescence ± spreading pubescent. Sepals connate for up to 2 mm. from base, 2–2·4–3 mm. long, pubescent. Petals 5·5–6·5–8 mm. long and 1·5–1·9–2·5 mm. broad. Capsules as var. *viridiflorum*. Fig. 1/2, p. 6.

TANGANYIKA. E. Usambara Mts., Uberi–Monga, 26 Jan. 1939, *Greenway* 5827!; Tanga District: Magunga Estate, 12 Dec. 1952, *Faulkner* 1093!; Rungwe District: Mbaka Forest, 30 Aug. 1912, *Stolz* 1542!
DISTR. T3,7; Malawi, Mozambique, Rhodesia, South Africa (Transvaal, Natal, Cape Province) and Angola
HAB. Lowland and upland rain-forest, particularly margins, rocky slopes and secondary growth; 450–2000 m.

SYN. *P. malosanum* Baker in K.B. 1897: 244 (1897); Dur. & Schinz, Consp. Fl. Afr. 1 (2): 228 (1898); Engl. in V.E. 3 (1): 851 (1915); T.T.C.L.: 452 (1949), pro parte
P. antunesii Engl. in E.J. 32: 130 (1902). Type: Angola, Huila, *Antunes* 116 (BM, iso. !)
P. viridiflorum Sims subsp. *malosanum* (Baker) Cuf. in F.R. 55: 71, fig. 4/f, g & 5/a (1952)

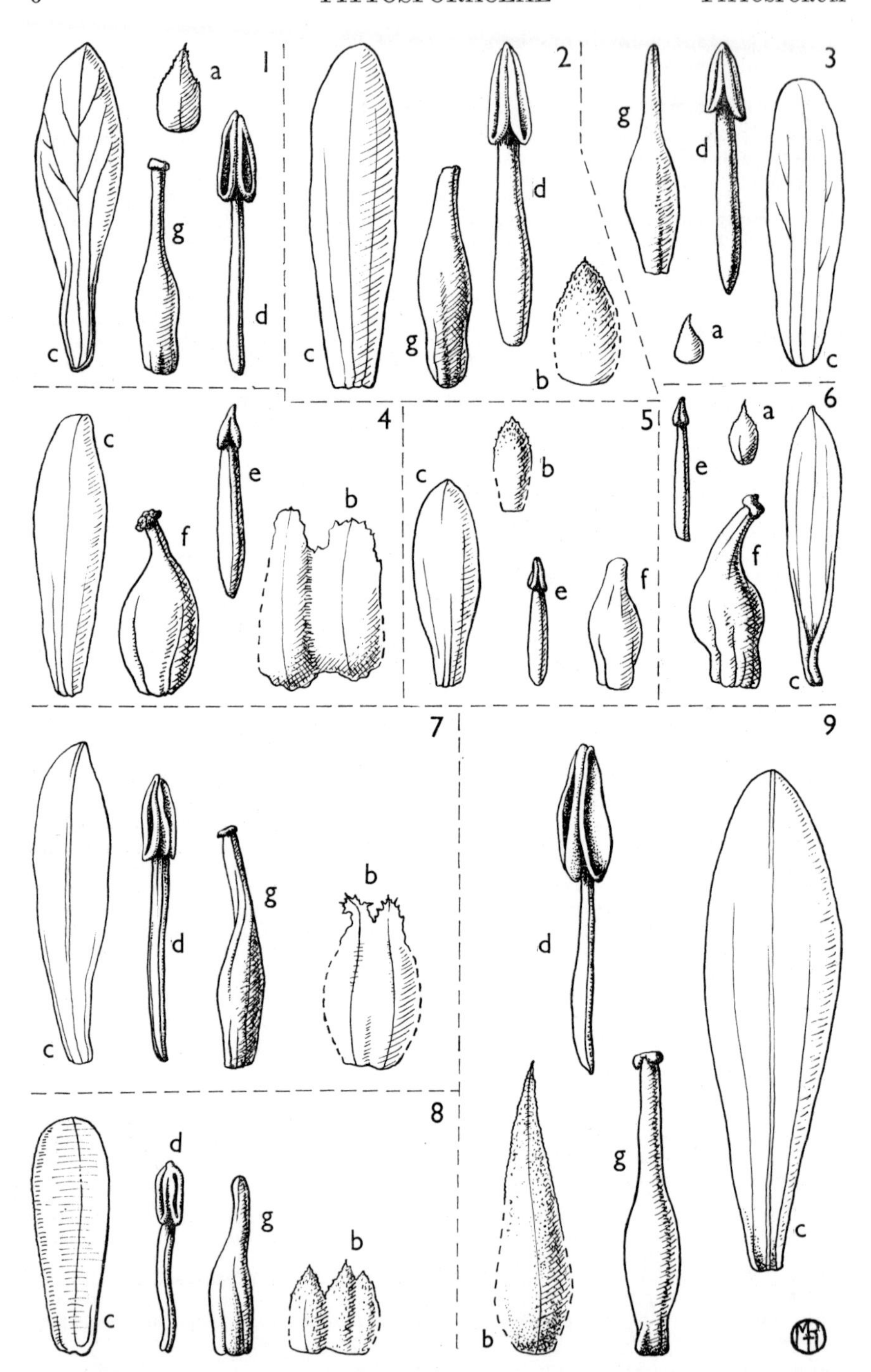

FIG. 1. ***PITTOSPORUM*—a,** free sepals; **b,** parts of connate calyx; **c,** petals; **d,** fertile stamens; **e,** sterile stamens; **f,** fertile gynoecium; **g,** sterile gynoecium of various species, all × 6. **1,** *P. viridiflorum* var. *viridiflorum*; **2,** *P. viridiflorum* var. *malosanum*; **3,** *P. viridiflorum* var. *afrorientale*; **4,** *P. viridiflorum* var. *kruegeri*; **5,** *P. mannii* subsp. *mannii*; **6,** *P. mannii* subsp. *ripicola*; **7,** *P. spathicalyx*; **8,** *P. lynesii*; **9,** *P. undulatum*. 1, from *Teague* 322; 2, from *Faulkner* 1093; 3, from *Fries* 928; 4, from *Gardner* in *F.D.* 2398; 5, from *Brenan* 9398; 6, from *Hancock & Chandler* 55; 7, from *Kakaire* 155; 8, from *Gilchrist* 69; 9, from *Brain* 6095.

[*P. viridiflorum* sensu Brenan in Mem. N.Y. Bot. Gard. 8: 219 (1953) et auct. al. pro parte, *non* Sims sensu stricto]

NOTE. Apparently most frequent in Malawi, Rhodesia and the adjacent parts of Mozambique, and not recorded north of Tanganyika. It links var. *viridiflorum* with var. *kruegeri*, at least in the extratropical part of its range. It also resembles *P. spathicalyx* De Wild., which has, however, certain exclusive features and geographically is quite disjunct.

var. **afrorientale** (*Cuf.*) *Cuf.* in Bol. Soc. Brot., sér. 2, 34:164 (1960). Type: Mt. Kenya, W. slopes, Coles Mill, *Fries* 928 (UPS, holo. !, BR, K, iso. !)

Adult leaves ± as var. *viridiflorum*, but narrower with breadth-length ratio 1:2–1:4, quite glabrous or nearly so. Pedicels 4–6·7–10 mm. long, like all branches of inflorescence quite glabrous or with sparse scattered adpressed puberulence. Sepals free or slightly connivent at base, 1–1·6–2·5 mm. long, glabrous. Petals 7–7·4–8·5 mm. long and 1·5–1·8–2·5 mm. broad. Ripe capsule probably always more than 4-seeded; valves thickish, dorsally flat-subconvex, generally spreading horizontally. Fig. 1/3.

KENYA. Nakuru District: Thomsons Falls, 14 Apr. 1956, *Verdcourt* 1477 !; N. Nyeri District: W. slopes of Mt. Kenya, Coles Mill, 17 Jan. 1922, *Fries* 928 ! & N. of Nanyuki, 8 July 1938, *Pole Evans & Erens* 1217 !
TANGANYIKA. Mbulu District: Ngorongoro, 12 Nov. 1932, *Geilinger* 3641!; W. Usambara Mts., near Golgolo, 'Worlds View', 4 June 1953, *Drummond & Hemsley* 2843 !; Njombe District: Njombe–Mbeya, 9 Sept. 1956, *Semsei* 2463 !
DISTR. **K**3,4; **T**2,3,7; probably also Rwanda Republic
HAB. Upland rain-forest, dry evergreen forest and upland evergreen bushland, also riverine forest; (900–)1450–2550 m.

SYN. [*P. abyssinicum* sensu Oliv. in Trans. Linn. Soc., ser. 2, 2: 328 (1887); Dur. & Schinz, Consp. Fl. Afr. 1 (2): 227 (1898), pro parte, quoad *Meyer* 151; A. Chev. in Bull. Soc. Bot. Fr. 93: 206 (1946), pro parte, quoad *Meyer* 151, *Stolz* 2477; T.T.C.L.: 452 (1949) et auct. al. pro parte, *non* Del.]
[*P. vosseleri* sensu Fries in N.B.G.B. 8: 555 (1923); Engl. in V.E. 5 (1): 231 (1925) et auct. al. pro parte, *non* Engl.]
P. viridiflorum Sims subsp. *afrorientale* Cuf. in F.R. 55: 63, fig. 4/j (1952)

NOTE. This variety evidently arose from var. *viridiflorum* by a ± complete loss of indumentum, but seems to have attained a fairly marked taxonomic differentiation in characters of the fruit. It is apparently restricted to Tanganyika and Kenya, although *Troupin* 3101, a specimen with very young fruits from Rwanda Republic, possibly belongs here too. Flowers are necessary for reliable distinction from var. *kruegeri* and fruiting specimens from N. Tanganyika are referable here, therefore, only with reserve.

var. **kruegeri** (*Engl.*) *Engl.* in V.E. 3 (1): 850 (1915); Cuf. in Bol. Soc. Brot., sér. 2, 34: 165 (1960). Type: South Africa, Transvaal, near Belfast, *Wilms* 213 (B, holo. †, BM, WU, iso. !)

Adult leaves as var. *viridiflorum*, with breadth-length ratio 1:2·5–1:4·3, quite glabrous. Pedicels 4–5·9–9 mm. long, like branches of inflorescence quite glabrous or nearly so. Calyx 2–3·8(–6) mm. long with sepals connate for 1·5–2·8(–4·5) mm., glabrous. Petals 6–7–8 mm. long and 2 mm. broad. Ripe capsule always (at least in East Africa) more than 4-seeded, similar to those of var. *afrorientale*. Fig. 1/4.

UGANDA. Karamoja District: Moruongole, June 1942, *Dale* 248 ! & Napak, 28 May 1940, *A. S. Thomas* 3640 !; Elgon, Bukwa–Kyosoweri, 16 Apr. 1927, *Snowden* 1069 !
KENYA. Elgon, Mar. 1931, *Lugard* 627 ! ; Nakuru, *Scott Elliot* 182 ! ; Naivasha/Kiambu District: [Kedong] Escarpment, near Katama stream, Aug. 1930, *Gardner* in *F.D.* 2398 !
TANGANYIKA. Bukoba District: Karagwe, Sept.–Oct. 1935, *Gillman* 594 !; probably also Mwanza District: Ruande Forest Reserve, Mdandawamungu R., 3 July 1952, *Gane* 32 !
DISTR. **U**1,3 ; **K**3–7 ; **T**1 ; W. Somali Republic (N.), French Somaliland, Ethiopia, probably southern Sudan Republic, also Rhodesia and South Africa (Transvaal, Swaziland, Orange Free State, Natal and Basutoland)
HAB. Upland dry evergreen forest, riverine forest and wooded grassland; 900–2400 m.

SYN. *P. kruegeri* Engl. in N.B.G.B. 2: 26 (1897)
[*P. abyssinicum* sensu Fries in N.B.G.B. 8: 555 (1923) et auct. mult. quoad Aethiopiam etc., pro parte maxima, *non* Del.]
P. quartinianum Cuf. in Oest. Bot. Zeitschr. 98: 132, fig. 56 (1951); I.T.U., ed. 2: 314 (1952). Type: Eritrea, Mt. Lesa, *Pappi* (216) 4660 (FI, holo. !, K, W, iso. !)

[*P. viridiflorum* Sims subsp. *somalense* sensu Cuf. in F.R. 55: 64 (1952), pro parte, tantum quoad fig. 4/k, l, *non* Cuf. sensu stricto]
P. viridiflorum Sims subsp. *kruegeri* (Engl.) Cuf. in F.R. 55: 68, fig. 4/q–s (1952)
P. viridiflorum Sims subsp. *quartinianum* (Cuf.) Cuf. in F.R. 55: 69, fig. 4/t–w (1952) & E.P.A.: 174 (1954); K.T.S.: 382, fig. 74/a, c, d (1961)

NOTE. It seemed advisable to unite *P. kruegeri* and *P. quartinianum* despite the fact that the latter is characterized by constantly spreading thick ripe valves and ranges from Tanganyika northwards, while the former seems to be variable in fruit-form and occurs from Rhodesia southwards. Var. *kruegeri* may perhaps actually be, therefore, diphyletic, but much more and especially fruiting material is required to establish its true status. Flowering specimens are easily separated from var. *malosanum* by the complete or almost complete lack of indumentum and from var. *afrorientale* by the connate sepals, but fruiting ones are hardly distinguishable from the latter.
All specimens collected from the Karamoja District of Uganda show long connate calyces perfectly resembling those of *P. spathicalyx* De Wild. in shape, but quite glabrous (like the rest of the plant). They are interpreted, therefore, as extreme developed forms of var. *kruegeri*. See also note under *P. spathicalyx* (p. 10).

2. **P. mannii** *Hook. f.* in J.L.S. 6: 5 (1862); Z.A.E. 2: 183 (1922); Cuf. in F.R. 55: 81, fig. 5/f–h & 8 (1952) & in F.W.T.A., ed. 2, 1: 182 (1954) & in F.Z. 1: 300, t. 54/D (1960) & in Bol. Soc. Brot., sér. 2, 34: 167 (1960), et auct. quoad Cameroon Mt. et Fernando Po. Type: Fernando Po, Clarence Peak, *Mann* 640 (K, holo. !)

Shrub or tree, 2–5·5–12 m. high; bole up to 40 cm. in diameter, with rough grey bark. Leaves mostly crowded at ends of flowering branches, with blade obovate, broadly oblanceolate or spathulate, together with petiole up to 17·5 cm. long and 4 cm. broad (average 12·5 × 3·7 cm.), always ± acuminate but usually with bluntish tip, tapering into the up to 20 mm. long petiole, ± deep green and sometimes glossy above, paler and often yellowish-green beneath, quite glabrous or very rarely somewhat puberulous on the lower parts when young; midrib considerably impressed in a ± deep groove above and often almost hidden, prominent beneath; lateral nerves up to 8 on either side, interconnected by loops; reticulation lax, not uniform, looser along the midrib, forming a ± distinct areolation above, dark coloured later turning pale and prominulous like the nerves beneath. Inflorescences terminal racemose panicles, mostly many-flowered, often with additional subumbellate axillary branches at the base; axes all ± pubescent; bracts minute, subulate, pubescent, early caducous. Flowers strongly and agreeably sweet-scented, white, cream, yellowish or greenish, on 3–10 mm. long pubescent pedicels. Sepals lanceolate or linear-lanceolate, free, not imbricate and deciduous or slightly connate and persistent, 1–3 mm. long and up to 1 mm. broad, mostly puberulous or ciliolate. Petals sublinear, 3·5–7 mm. long and 1·5–2 mm. broad, erect or recurved. Fertile stamens 3–6·5 mm. long with 1–2·5 mm. long anthers; sterile stamens 3–4 mm. long with ± 0·75 mm. long anthers. Gynoecium 3–5 mm. long if fertile, 3·5–5·5 mm. long if sterile, with ovary sometimes slightly puberulous. Capsules never maturing more than 4 seeds; valves of ripe capsule thin, dorsally concave and darker, finally ± reflexed.

subsp. **ripicola** (*J. Léon*) *Cuf.* in F.Z. 1: 302, t. 54/D (1960) & in Bol. Soc. Brot., sér. 2, 34: 168 (1960); K.T.S.: 380 (1961). Type: Congo Republic, Ubangi-Uele, Dakwa on R. Dolo, *Germain* 718 (BR, holo. !)

Lateral nerves and reticulation of old leaves turning pale but mostly very little or not at all prominulous beneath. Pedicels 3–5·8–10 mm. long. Sepals free, deciduous, up to 3 × 1 mm. (average 1·5 mm. long). Petals 4·5–5·9–7 mm. long, erect with slightly spreading tips. Fertile stamens 4–5·2–6·5 mm. long with 1–1·6–2·5 mm. long anthers; sterile stamens ± 3·5 mm. long with 0·75 mm. long anthers. Fertile gynoecium ± 4·75 mm. long; sterile gynoecium ± 4·5 mm. long. Fig. 1/6, p. 6 & 2/8,9, p. 11.

UGANDA. Busoga District: Bunya, Nov. 1937, *Webb* 46 !; Mengo District: Nambigiluwa

[Nambigiruwa] Swamp, Jan. 1932, *Eggeling* 155 in *F.H.* 367 ! & 19 km. on Kampala–Entebbe road, Apr. 1932, *Eggeling* 395 in *F.H.* 677 !
KENYA. Trans-Nzoia District: Kitale, June 1942, *Webster* in *Herb. Amani* 8741 !; N. Kavirondo District: Kakamega, June 1933, *Dale* in *F.D.* 3081 !; Kericho, Sept. 1933, *Napier* 5356 !
TANGANYIKA. Ngara District: Bushubi, Keza, 20 May, 1960, *Tanner* 4938 !; Kigoma District: Mahali Mts., Sisaga, 29 Aug. 1958, *Newbould & Jefford* 1923 !
DISTR. **U**1–4; **K**3,5; **T**1,4; Ethiopia, Sudan Republic (White Nile District), NE. and SE. Congo Republic, Rhodesia, Central African and Cameroun Republics, S. Nigeria and probably extending into Dahomey, Ivory Coast and Guinée Republic
HAB. Rain-forest, particularly at margins, riverine, lakeside and swamp-edge forest, rarely in bamboo thicket or wooded grassland; 1140–1650(–2100) m.

SYN. [*P. abyssinicum* sensu auct. pro parte, quoad Aethiopiam, Sudan, Uganda, *non* Del.]
[*P. mannii* sensu auct. pro parte, exceptis Fernando Po et Mt. Cameroon, *non* Hook. f. sensu stricto]
[*P. fragrantissimum* sensu De Wild., Pl. Bequaert. 2: 47 (1923), pro parte max., excepto specim. *Battiscombe* 644, *non* Engl.]
[*P. dalzielii* sensu A. Chev. in Bull. Soc. Bot. Fr. 93: 206 (1946), pro parte max. excepto specim. *Dalziel* 417, *non* Hutch.]
P. ripicola J. Léon. (as "*ripicolum*") in B.J.B.B. 20: 47, fig. 7 (1950); Cuf. in Oest. Bot. Zeitschr. 98: 129 (1951); J. Léon. in F.C.B. 2: 579, fig. 9 (1951); Cuf. in F.R. 55: 77, fig. 5/d,e & 8 (1952); I.T.U., ed. 2: 314 (1952); Tisserant & Sillans in Not. Syst. 15: 92 (1954); E.P.A.: 173 (1954); F.F.N.R.: 66 (1962)
P. ripicola J. Léon. subsp. *katangense* J. Léon. in B.J.B.B. 20: 227 (1950) & in F.C.B. 2: 580 (1951) prob. huc

NOTE. Subsp. *mannii* (fig. 1/5) seems to be restricted to Fernando Po and the Cameroun Republic (Cameroon Mt., Bamenda and the adjacent higher parts of the Bambutos). It differs from subsp. *ripicola* mainly in the dimensions of the flowers, of which the main features and average measurements are as follows: pedicels 4·9 mm. long; sepals connate at the base and persistent, 1·8 mm. long; petals strongly reflexed, 4·8 × 1·7 mm.; fertile stamens 3·6 mm. long with 1·3 mm. long anthers; sterile stamens (like subsp. *ripicola*) 3·5 mm. long; fertile gynoecium 3 mm. long; sterile gynoecium 3·5 mm. long. Some difference in ecological requirements must be associated with the noticeably higher altitude records of 915–2600 m. with an average of 1880 m. In the absence of flowers the two subspecies cannot be reliably distinguished, and for this reason all the specimens collected from west of Nigeria cannot be certainly placed.

P. mannii replaces *P. viridiflorum* in the middle and western parts of northern tropical Africa and these closely related species overlap only in Uganda, Kenya and Ethiopia.

3. **P. spathicalyx** *De Wild.*, Pl. Bequaert. 2: 45 (1923); Lebrun, Les Ess. For. Congo Orient.: 81 (1935); F.P.N.A. 1: 232, t. 22 (1948), J. Léon. in F.C.B. 2: 576, t. 57 (1951); Cuf. in F.R. 55: 74, fig. 5/b, c & 8 (1952); I.T.U., ed. 2: 314 (1952); Cuf. in Bol. Soc. Brot., sér. 2, 34: 167 (1960). Type: Congo Republic, near Lake Edward, Angi, *Bequaert* 5826 (BR, holo. !)

Shrub or small tree, 3–13 m. high; bole up to 15 cm. in diameter. Young branches hispidulous-pubescent. Leaves not crowded at ends of branches, with blade broadly lanceolate to obovate, together with petiole 7–11·5–15 cm. long, 2–3–4·5 cm. broad, ± acuminate, bluntish, tapering to an up to 15 mm. long petiole, dull green above, paler beneath, ± hispidulous-pubescent later glabrescent; midrib rather deeply impressed above, prominent beneath; lateral nerves up to 12 on either side, unequal, prominulous, ± interconnected; reticulation dense, uniform, dark coloured beneath and colour disappearing only at a very late stage. Inflorescences terminal slender racemose panicles; all branches spreading ferrugineous pubescent; bracts minute, subulate, pubescent, early caducous. Flowers sweet-scented like *Jasminum* or *Narcissus*, white, cream, yellowish or greenish, on 3–5–8 mm. long densely ferrugineous pubescent pedicels. Calyx 3–3·3–4 mm. long, connate for 2–2·5–3 mm., enveloping the buds, early split down one side; lobes short, blunt, ciliolate. Petals linear, 5–6·5–8 mm. long, 1·5–1·75–2 mm. broad, usually strongly reflexed. Fertile stamens 4–5·6–7 mm.

long with 1·5–1·8–2 mm. long anthers; sterile stamens 3 mm. long with 0·75 mm. long anthers. Fertile gynoecium with short stipe and style 3·5 mm. long, scanty puberulous; sterile gynoecium 3–5–7 mm. long. Capsules (?4–)6–8-seeded; valves of ripe capsule up to 10 mm. in diameter, uniformly thin, dorsally flat-convex to weakly gibbous, finally spreading or somewhat reflexed. Fig. 1/7, p. 6.

UGANDA. Ankole District: Bugamba Forest, June 1939, *Cree* 219 !; Kigezi District: Rubaya, July 1946, *Purseglove* 2093 ! & Kachwekano Farm, July 1949, *Purseglove* 2975 !

KENYA. Baringo District: Kamasia, Katimok Forest, Oct. 1930, *Dale* in *F.D.* 2408 !

TANGANYIKA. Bukoba District: Kiamawe Forest, Sept.–Oct. 1935, *Gillman* 459 ! & Nshamba, Sept.–Oct. 1935, *Gillman* 619 ! & Lushambia-Niabura, 26 June 1913, *Braun* in *Herb. Amani* 5564 !

DISTR. **U**2; **K**3; **T**1,4; eastern Congo, Rwanda and Burundi Republics

HAB. Upland rain-forest, particularly margins and near rock-outcrops, riverine forest and upland evergreen bushland; 1200–2400 m.

SYN. [*P. abyssinicum* sensu Baker, Moore & Rendle in J.L.S. 37: 123 (1905), *non* Del.]
[*P. mannii* sensu Engl. in V.E. 3 (1): 851 (1915), pro parte, quoad exsicc. *A. Braun*, *non* Hook. f.]
[*P. malosanum* sensu T.T.C.L.: 452 (1949), pro parte, quoad Bukoba District, *non* Baker]
[*P. viridiflorum* Sims subsp. *quartinianum* transit ad *P. spathicalyx* sensu Cuf. in F.R. 55: 49 (1952), quoad exsicc. *Gillman* 619 et *Dale* in *F.D.* 2408]

NOTE. *Dale* in *F.D.* 2408 (with very young fruits, see K.T.S. 381, fig. 74/b, c) is the only record for Kenya and is referred here with some hesitation. The locality is rather disjunct from the main area of the species. It is possibly related to divergent forms of *P. viridiflorum* var. *kruegeri* from Karamoja, briefly discussed under the latter (p. 8). Material to date is too scanty for a safe judgement. The two taxa seem otherwise to be in immediate contact only in the extreme north-western part of Tanganyika adjacent to the Uganda boundary.

4. **P. lynesii** *Cuf.* in F.R. 55: 84, fig. 5/k,l & 8 (1952) & in Bol. Soc. Brot., sér. 2, 34: 168 (1960). Type: Tanganyika, Njombe, *Lynes* V/75 (K, holo.!)

Tree 6–18 m. high, with trunk up to 45 cm. in diameter and often rather stout in proportion to height; bark reddish or brown, fissured. Leaves clustered at apices of flowering branches, with blade broadly lanceolate to elliptic, together with petiole 9·5–10–12·5 cm. long and 3·5–4 cm. broad, rounded, blunt or shortly and bluntly acuminate, ± long-cuneate into the up to 15 mm. long petiole, dull green, much paler beneath, quite glabrous from inception; midrib rather deeply impressed above, markedly prominent beneath; lateral nerves up to 11 on either side, very thin, irregularly interconnected; reticulation dense, uniform, delicately areolate above, clearly visible by dark persistent colour beneath. Inflorescences terminal racemose panicles, with subverticillate branches, up to 5 cm. long, sometimes with subumbellate accessory branches at the base, ± densely clothed overall by rusty or coppery, somewhat shiny, spreading pubescence; bracts lanceolate, up to 2 mm. long and 1 mm. broad, glabrous except the ciliolate edges, caducous. Flowers strongly and sweetly scented, whitish, cream or yellowish, on 4–5·5–8 mm. long pubescent pedicels. Sepals ovate, ± 2 mm. long and 1 mm. broad, rounded, connate sometimes up to the middle, quite glabrous, deciduous all together. Petals distinctly spathulate, 5–6–7 mm. long, 1·5–2–2·5 mm. broad, strongly recurved from the middle. Fertile stamens 4·5–5·5–7 mm. long with ± 1·75 mm. long anthers; sterile stamens unknown. Gynoecium 4–5·5 mm. long if sterile (fertile unknown), nearly glabrous. Capsule more than 4-seeded; valves of ripe capsule up to 8 mm. in diameter, thickish, almost flat on both sides, spreading horizontally. Fig. 1/8, p. 6.

TANGANYIKA. Mbeya/Rungwe District: Poroto Mts., 16 Mar. 1932, *St. Clair-Thompson* 915 !; Iringa District: Mufindi, Kigogo Forest, 30 Nov. 1946, *Gilchrist* 69 in *F.H.* 1957 ! & 20 Jan. 1952, *Wigg* 1000 !

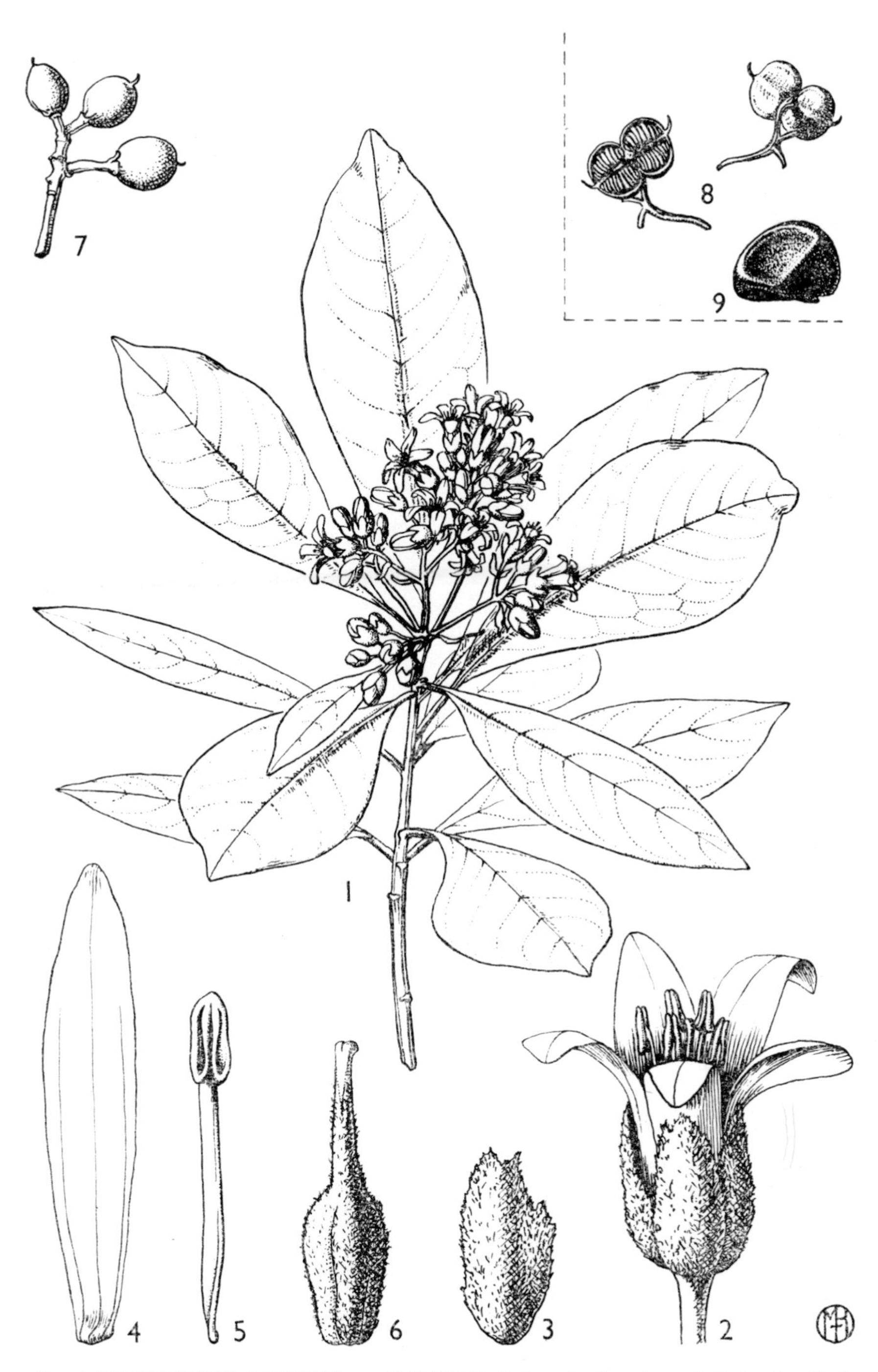

FIG. 2. *PITTOSPORUM LANATUM* var. *LANATUM*—**1,** flowering branch, × 1; **2,** functionally ♂ flower, × 5; **3,** sepal × 6; **4,** petal, × 6; **5,** fertile stamen, × 6; **6,** sterile gynoecium, × 6; **7,** young capsules, × 1. *P. MANNII* subsp. *RIPICOLA*—**8,** dehisced capsule, × 1; **9,** seed, × 4. 1–6 from *Battiscombe* 908; 7, from *Munro* 108; 8,9, from *Lugard* 627.

DISTR. **T**7; not known elsewhere
HAB. Upland rain-forest and bamboo thicket, also regenerating bushland; 1830–2400 m.

SYN. [*P. malosanum* sensu T.T.C.L.: 452 (1949), pro parte, quoad specim. *Wigg* 27, *non* Baker]

NOTE. The individual characters of this species are not very striking, yet their combination is not met with elsewhere. It seems most akin to and presumably a local derivative of *P. viridiflorum* in the broad sense.

5. **P. lanatum** *Hutch.* & *Bruce* in K.B. 1941: 97 (1941); I.T.U., ed. 2: 313 (1952); Cuf. in Bol. Soc. Brot., sér. 2, 34: 171 (1960). Type: Ethiopia, Harar, Mt. Kondudo, *Gillett* 5228 (K, holo.!)

Tree 3·5–8·5–18·5 m. high; bole 20–30 cm. in diameter. Flowering branches rusty pubescent. Leaves crowded at ends of branches, with blade obovate to oblanceolate, with petiole 6–8·3–10 cm long, 1·8–2·5–3·5 cm. broad, rounded to shortly acuminate, blunt, tapering into the up to 10 mm. long petiole, dull dark green above, paler beneath, finally glabrous above, persistently fulvous-tomentose beneath; midrib ± flat above, prominent beneath; lateral nerves thin, scarcely visible above, hidden by the tomentum beneath; reticulation dense, uniform, poorly visible above, hidden by the tomentum beneath. Inflorescences racemose panicles; all branches rusty pubescent; bracts lanceolate to subulate, 4–5–6 mm. long and 2 mm. broad at base, glabrous above, pubescent beneath and on margins, caducous. Flowers scented, bright to dull yellow, on 1·5–3–6 mm. long densely rusty pubescent pedicels. Sepals free, imbricate, bluntly ovate, 3–3·8–4·5 mm. long and 1–2–3 mm. broad, glabrous inside, ± pubescent outside. Petals (5–)6·5–8·4–10 mm. long and 2–2·5 mm. broad, ciliolate. Fertile stamens 5·5–6·3–7·5 mm. long with 1·5–2 mm. long anthers; sterile stamens 5 mm. long with 1 mm. long anthers. Gynoecium ± 5 mm. long if fertile, 5–5·8–6·5 mm. long if sterile, with densely tomentellous ovary and glabrous style. Capsule usually more than 4-seeded; valves of ripe capsule ovate, very thickened at base, dorsally gibbous, thinner at edges, suberect-divergent, glabrescent.

var. **lanatum**

Leaf-blade obovate-oblong, rounded or bluntly pointed at apex; petals 6·5–10 mm. long. Fig. 2/1–7, p. 11.

UGANDA. Karamoja District: Mt. Moruongole, 11 Nov. 1939, *A. S. Thomas* 3305! & June 1942, *Dale* U249 ! & Mt. Kadam, *J. Wilson* 762 !
KENYA. Elgeyo District: Marakwet Hills, Feb. 1934, *Dale* K672 !; Nakuru District: Mau, *Battiscombe* 957 ! & Eastern Mau Forest Reserve, 27 Aug. 1949, *Maas Geesteranus* 5922 !
TANGANYIKA. Mbeya District: Kikondo, 20 Oct. 1956, *Richards* 6644 !; Rungwe Mt., Nov. 1931, *R. M. Davies* R/21 ! & Oct. 1959, *Procter* 1454 !
DISTR. **U**1; **K**3,4; **T**7; Ethiopia
HAB. Upland rain-forest and dry evergreen forest, often at margins or in isolated forest-clumps, also upland grassland; 2250–2850 m.

SYN. [*P. fragrantissimum* sensu De Wild., Pl. Bequaert. 2: 47 (1923), pro parte. quoad specim. *Battiscombe* 644, *non* Engl.]
[*P. abyssinicum* sensu T.S.K.: 20 (1936), *non* Del.]
[*P. fulvo-tomentosum* sensu T.T.C.L.: 452 (1949), *non* Engl.]
P. abyssinicum Del. subsp. *lanatum* (Hutch. & Bruce) Cuf. in F.R. 55: 96, fig. 7/a,b (1952) & E.P.A.: 173 (1954); K.T.S.: 380 (1961)

NOTE. Only one gathering of this species, *Gardner* 428 from Kinale in the Kiambu District of Kenya, bears ripe fruits, on which the above description of these is based but what variation occurs over the range as a whole is not known. *P. lanatum*, together with *P. fulvo-tomentosum* Engl. (on the Virunga Mts. of the Congo/Rwanda border region) and *P. abyssinicum* Del. (restricted to Ethiopia) comprise a complex of closely related taxa, which it seems preferable to keep specifically separate for the time being.

Var. *engleri* (Cuf.) Cuf. (*P. tomentosum* Engl. *non* Bonpl., *nom. illegit.*; *P. engleri* Cuf.) is based on *Ellenbeck* 1898 from Ethiopia, a specimen no longer in existence, and according to the description, cannot represent more than a narrow-leaved small-flowered form of *P. lanatum*.

6. **P. goetzei** *Engl.* in E.J. 28: 392 (1900) & in V.E. 1 (1): 364 (1910) & 3 (1): 851 (1915); T.T.C.L.: 452 (1949); Cuf. in F.R. 55 : 86, fig. 5/m & 8 (1952) & in Bol. Soc. Brot., sér. 2, 34 : 169 (1960). Type: Tanganyika, Uluguru Mts., Lukwangule Plateau, *Goetze* 277 (B, holo. †); Uluguru Mts., *Lommels* 1068 (EA, neo. !)

Tree up to 8 m. high with rounded crown and grey bark. Branches hairy and densely leafy. Leaves with blade broadly oblanceolate to obovate, together with petiole up to 1·5–6·5 cm. long and 1–2·7 cm. broad, blunt at apex, sessile or narrowed into the up to 3 mm. long petiole, dull green, paler beneath, shortly patent-hairy, soon glabrescent except on the petiole; midrib and ± 12 alternately stronger and weaker lateral nerves on either side slightly impressed above, the latter flat but relatively broad beneath; reticulation dense, uniform, delicately engraved above, dark coloured beneath and finally turning pale but hardly prominulous. Inflorescences terminal, subracemose, few-flowered, here and there with 2-flowered branches, rather densely covered overall with short patent hairs; bracts subulate, up to 6 mm. long, ciliolate. Flowers yellowish-white on 10–18 mm. long hairy pedicels. Sepals free, not imbricate, ovate-lanceolate, 4–6 mm. long and 1·5–2·5 mm. broad, subacute, sparsely hairy. Petals sublinear, 10–14 mm. long and 3 mm. broad, erect. Fertile stamens 6–7 mm. long with 2 mm. long anthers; sterile stamens unknown. Gynoecium up to 7 mm. long if sterile (fertile unknown), with slightly hairy ovary. Capsules shortly stipitate, more than 4-seeded; valves of completely ripe capsule unknown, but valves of just opened capsule broadly ovate, ± 13 mm. long, rugulose, glabrescent except at base.

TANGANYIKA. Uluguru Mts., Oct.–Nov. 1905, *Lommels* 1068 ! & S. Uluguru Forest Reserve, Mar. 1955, *Semsei* 2071 !

DISTR. **T6**; known only from the Uluguru Mts.

HAB. Upland rain-forest, reported to be very common and the principal component above the belt of bamboo thicket; ± 2400 m.

SYN. [*P. malosanum* sensu A. Chev. in Bull. Soc. Bot. Fr. 93: 205 (1946), *non* Baker]

NOTE. An evidently very isolated and still poorly known species, perhaps most closely related to *P. mildbraedii* Engl., which is endemic in the Virunga Mts.

INDEX TO PITTOSPORACEAE

GEOGRAPHICAL DIVISIONS OF THE FLORA

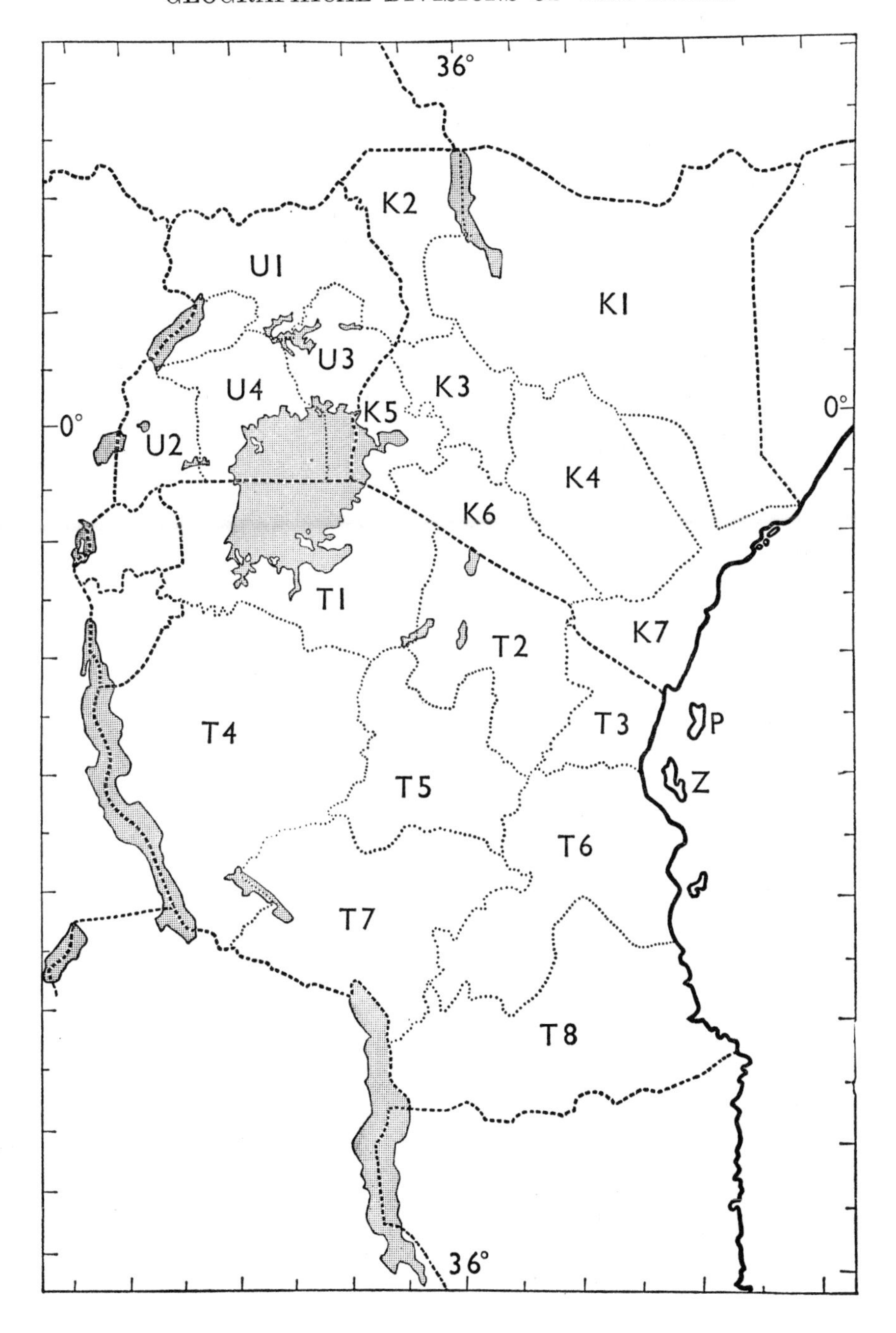